CIALIS
MEN SEXUAL HEALTH (TADALAFIL)

How To Clear Erectile Dysfunction, Undrestand Men & Women Sexual Health And Stay Healthy Using Cialis Tadalafil

Emilia Jax

Contents

Chapter 1.............................3

  Overview of cialis 3

Chapter 2............................6

  Uses of cialis.......................6

Chapter 3...........................15

  Precautions and warnings of cialis 15

Chapter 4..........................38

  Dosage of cialis.................38

The end 48

Chapter 1

Overview of cialis

Cialis, known generically as tadalafil, is a medication primarily used to treat erectile dysfunction (ED) in men. It belongs to a class of drugs called phosphodiesterase type 5 (PDE5) inhibitors.

Mechanism of Action:

Cialis works by inhibiting the enzyme PDE5, which regulates blood flow in the penis. This inhibition allows for increased blood flow to the penis during sexual stimulation, enabling an erection to occur naturally. Unlike some other ED medications, Cialis has a longer duration of action, which can last up to 36 hours, earning it the nickname "the weekend pill."

Chapter 2

Uses of cialis

1. Erectile Dysfunction (ED):

- **Purpose**: Cialis is primarily prescribed to treat erectile dysfunction (impotence) in men.

- **Mechanism**: It works by increasing

blood flow to the penis during sexual stimulation, enabling an erection.

- **Dosage**: Typically taken as needed, about 30 minutes to 1 hour before sexual activity. The effects can last up to 36 hours, earning it the nickname "the weekend pill."

2. Benign Prostatic Hyperplasia (BPH):

- **Purpose**: Cialis is also approved to treat symptoms of benign prostatic hyperplasia (BPH), which is an enlargement of the prostate gland.

- **Mechanism**: It helps relax the smooth muscles in the prostate and bladder, which can improve urinary symptoms such as difficulty urinating, urgency, or frequency.

- **Dosage**: Usually taken once daily at a lower dose, regardless of sexual activity.

3. Pulmonary Arterial Hypertension (PAH):

- **Brand Name**: Under the brand name Adcirca, tadalafil is used to treat pulmonary arterial hypertension (PAH).

- **Purpose**: It helps relax the blood vessels in the lungs, allowing blood to flow more easily and reducing the workload on the heart.

- **Dosage**: The dosage and frequency are determined by a healthcare provider based on individual patient needs.

Off-Label Uses:

While not FDA-approved for these uses, Cialis may be prescribed off-label for:

- **Raynaud's Phenomenon**: A condition where certain areas of the body, usually fingers and toes, feel numb and cold in response to cold temperatures or stress.

- **High-Altitude Pulmonary Edema**

(HAPE): A condition that can occur when ascending to high altitudes too quickly.

Chapter 3

Precautions and warnings of cialis General Precautions

1. **Allergies:**

    - Do not use Cialis if you are allergic to tadalafil or any of its components. Allergic reactions can include rash, itching, swelling, severe dizziness, and difficulty breathing.

2. **Medical History:**

   ◦ Inform your healthcare provider of your complete

medical history, especially if you have or have had heart problems (such as heart attack, irregular heartbeat, angina, heart failure), stroke, high or low blood pressure, liver or kidney disease, dehydration, eye problems (like retinitis pigmentosa, sudden vision loss), bleeding disorders, stomach ulcers, or conditions that may increase the risk of priapism (painful/prolonged erection) such as sickle cell anemia, leukemia, or multiple myeloma.

Drug Interactions

1. **Nitrates:**

- Cialis should not be used with nitrates (e.g., nitroglycerin, isosorbide dinitrate/mononitrate) commonly used for chest pain or heart problems. The combination can lead to a significant and potentially dangerous drop in blood pressure.

2. **Alpha-Blockers:**

    - Use with alpha-blockers (medications for high blood pressure or prostate conditions) can cause a drop in blood pressure, leading to dizziness or fainting. Patients should be stable on alpha-blocker therapy before starting Cialis, and the starting dose of Cialis should be low.

3. **Other Medications:**

   - Inform your doctor about all other medications you are taking, including certain antifungals (like ketoconazole, itraconazole), antibiotics (such as erythromycin), HIV protease inhibitors (e.g., ritonavir, saquinavir), hepatitis C virus protease inhibitors (e.g., boceprevir, telaprevir), and other medications for ED.

Specific Populations

1. **Cardiovascular Risk:**
   - Sexual activity carries a potential cardiac risk, especially

in patients with preexisting cardiovascular disease. Use Cialis with caution if you have heart-related issues and consult your healthcare provider to evaluate if you are fit enough for sexual activity.

2. **Vision Problems:**

   ○ Rarely, Cialis can cause sudden vision loss in one or both eyes (non-arteritic anterior ischemic optic neuropathy, NAION). If you experience sudden vision loss, stop taking Cialis and seek immediate medical attention.

3. **Hearing Problems:**

- Cialis can cause sudden decrease or loss of hearing, sometimes accompanied by ringing in the ears and dizziness. If this occurs, discontinue use and consult a healthcare provider immediately.

4. **Priapism:**

- Priapism is a prolonged and painful erection lasting more than 4 hours. If this occurs, seek immediate medical help to prevent permanent damage to the penis.

Lifestyle and Dietary Considerations

1. **Alcohol:**
   - Limit alcohol consumption as it can increase the risk of side

effects such as dizziness, headache, and low blood pressure.

2. **Grapefruit Juice:**

    - Avoid grapefruit and grapefruit juice while using Cialis as it can increase the levels of tadalafil in the blood, leading to increased side effects.

Administration and Overdose

1. **Dosing:**

   ◦ Take Cialis exactly as prescribed by

your healthcare provider. Do not take more than the recommended dose or use it more frequently than prescribed.

2. **Overdose:**

    ◦ In case of overdose, seek immediate

medical attention. Symptoms of overdose may include severe dizziness, fainting, or a prolonged erection.

Monitoring and Follow-up

1. **Regular Check-ups:**

    - Regular follow-up visits with your healthcare provider are important to monitor your response to Cialis and any potential side effects.

2. **Kidney and Liver Function:**

   o Patients with severe kidney or liver impairment may require dosage adjustments and should be monitored closely.

Chapter 4

Dosage of cialis Erectile Dysfunction (ED):

1. **As Needed**:

    - **Starting Dose**: 10 mg taken prior to anticipated sexual activity.

- **Adjustments**: Based on efficacy and tolerability, the dose can be increased to 20 mg or decreased to 5 mg.
- **Maximum Frequency**: Once per day, as needed. Take at

least 30 minutes before sexual activity.

- **Duration of Effect**: Can last up to 36 hours.

2. **Daily Use**:

- **Dose**: 2.5 mg taken at the same time every day, regardless of timing of sexual activity.

- **Adjustment**: May be increased to 5 mg daily based on individual efficacy and tolerability.

Benign Prostatic Hyperplasia (BPH):

1. **Daily Use**:

    - **Dose**: 5 mg taken at the same time every day.

    - **Combination with Finasteride**: When used in combination with finasteride for BPH, the recommended dose is 5 mg taken once daily for up to 26 weeks.

Erectile Dysfunction and Benign Prostatic Hyperplasia (ED + BPH):

1. **Daily Use**:

- **Dose**: 5 mg taken at the same time every day, regardless of timing of sexual activity.

Pulmonary Arterial Hypertension (PAH) (under the brand name Adcirca):

1. **Dose**: 40 mg (two 20 mg tablets) taken once daily.

       ◦ **Note**: For PAH, the dosage and administration are usually managed and adjusted by a healthcare provider.

General Guidelines:

- **Food**: Cialis can be taken with or without food.

- **Alcohol**: Limit alcohol consumption as it can increase the risk of side effects like dizziness and low blood pressure.

- **Missed Dose** (for daily use): If you

miss a dose, take it as soon as you remember. If it is close to the time of your next dose, skip the missed dose. Do not double the dose to catch up.

The end

Made in the USA
Las Vegas, NV
29 August 2024

94599948R10028